You Know You Live near a Factory Farm When Your Kids Go Fishing with a Pool Skimmer

Pure Profit Farms LLC

Mr. Factory Farm

An Uncensored Look at Today's Farming Practices
by Doug Baird

ISBN 978-0-9898608-7-1

2018

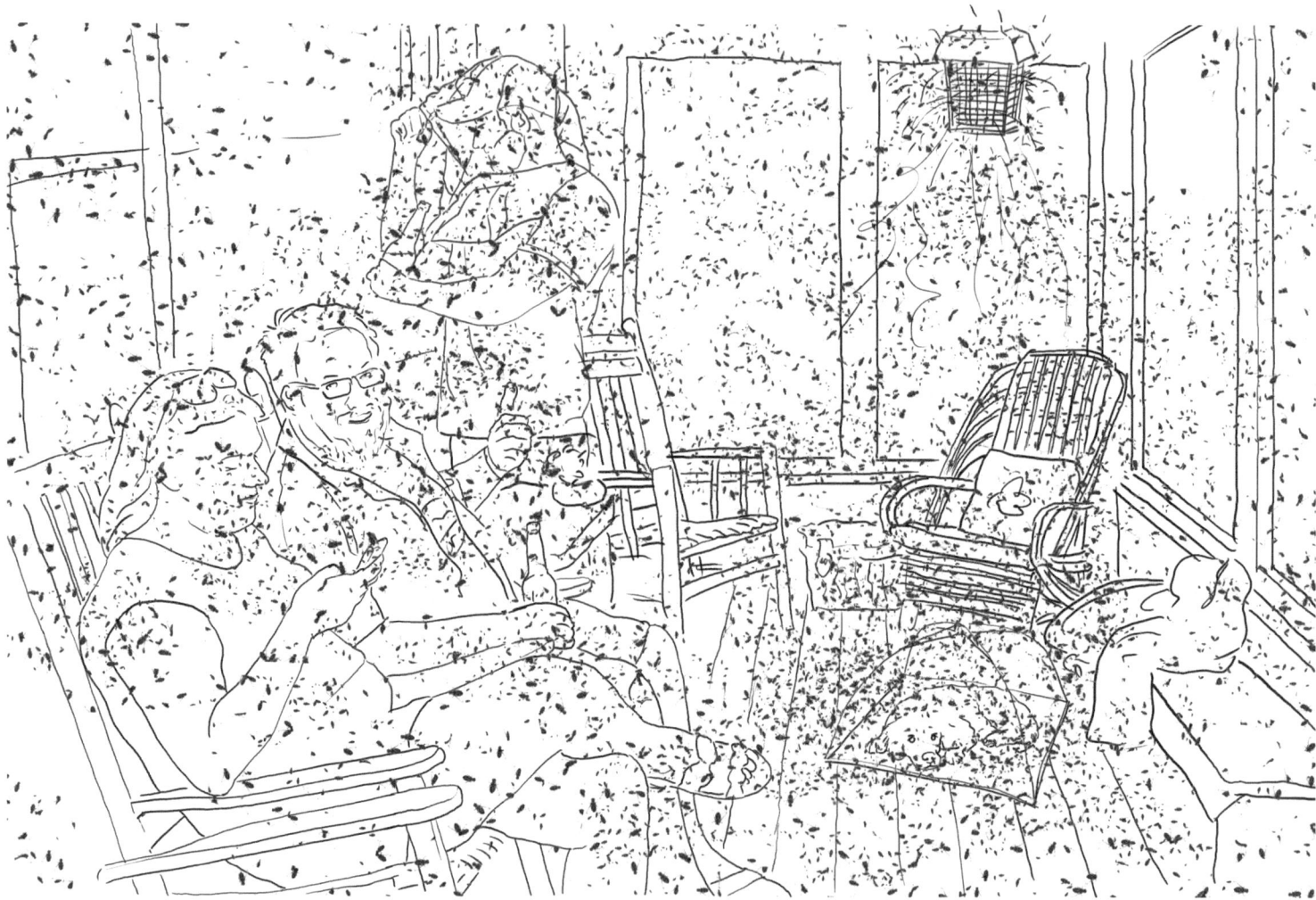

You know you live near a factory farm, when your bug-zapper sounds like a popcorn machine.

You know you live near a factory farm, when the morning mist is herbicide.

CAUTION:
HARMFUL ALGAL BLOOM ALERT
Avoid contact with blooms.

Keep people and pets away from blooms.
Harmful algal blooms have been seen in this waterbody.
Blooms can make you and your pets sick.
DO NOT EAT FISH CAUGHT IN A BLOOM AREA
LEARN MORE www.countygov.loveourlake.cleanup.org

DANGER
DO NOT DRINK
THIS WATER

You know you live near a factory farm, when the lake has more warning signs than tourists.

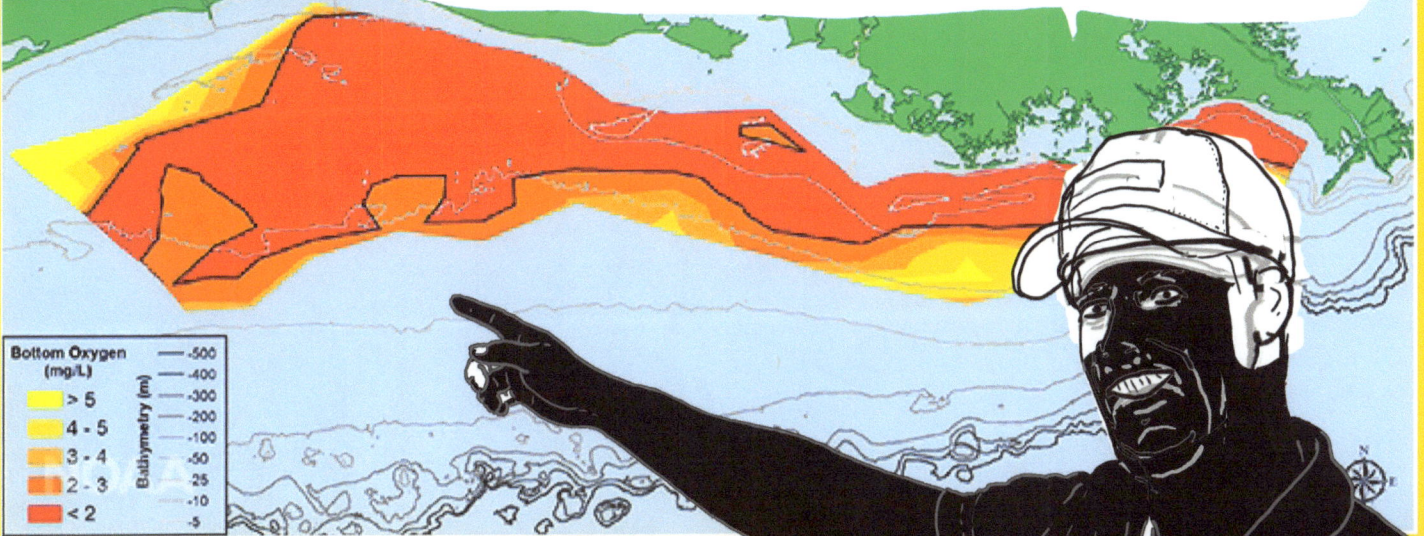

You know you live near a factory farm,
when a part of the Dead Zone
is named after it.

You know you live near a factory farm,
when manure
comes out of the tap.

You know you live near a factory farm,
when the farmers and regulators
are the same people.

You know you live near a factory farm,
when even your dog
has asthma.

This disclosure notice is to ... that the property they are about to buy lies partially or ... an agricultural district and that farming activities occur within the district. Such farming activities may include, but not limited to, activities that cause long-term respiratory problems, physical impairment, brain damage, cancer, blue baby syndrome, death, dust, noise and odors. Prospective residents are also informed that the location of property within an agricultural district may impact the ability to use and enjoy outdoor areas, and access potable well water, water and/or sewer services for such property under certain circumstances.

Such disclosure notice shall be signed by the prospective grantor and grant... prior to the sale, purchase or exchange of such real property.

Receipt of s... report form... for in sectio...

...notice shall be recorded on a property transfer ...vices as prov...

Don't worry — your safety's guaranteed under Agricultural Law. They've just blacked out some unnecessary information to avoid confusion.

You know you live near a factory farm, when the Real Estate Disclosure Form is redacted.

"**Pure Profit Farms** — USDA Certified Organic Milk"
From cows fed on: Hydrolyzed poultry feather meal, Manure solids and cattle manure, Processed poultry liver, Blood meal, Tallow, Ruminally inert fats, Meat and bone meal . . .

Mr. Factory Farm

You know you live near a factory farm, when USDA Certified means "you stupid, dumb assholes."

You know you live near a factory farm,
when your Bed and Breakfast
is shut and closed.

From the USDA National Agricultural Library:

What is an Invasive Species?

As per Executive Order 13112 an "invasive species" is defined as a species that is:

1) non-native (or alien) to the ecosystem under consideration and

2) whose introduction causes or is likely to cause economic or environmental harm or harm to human health.

You know you live near a factory farm, when you realize they're an invasive species.

REAL ESTATE AUCTION
House • Barn • 170+ acres
THU • MAY 3 • 11AM
Green Lakes
Auction
Company
ONE PURE PROFIT ROAD RURAL, NY (607) 486-6666
www.greenlakesauction.com

You know you live near a factory farm,
when it keeps getting bigger —
while your community gets smaller.

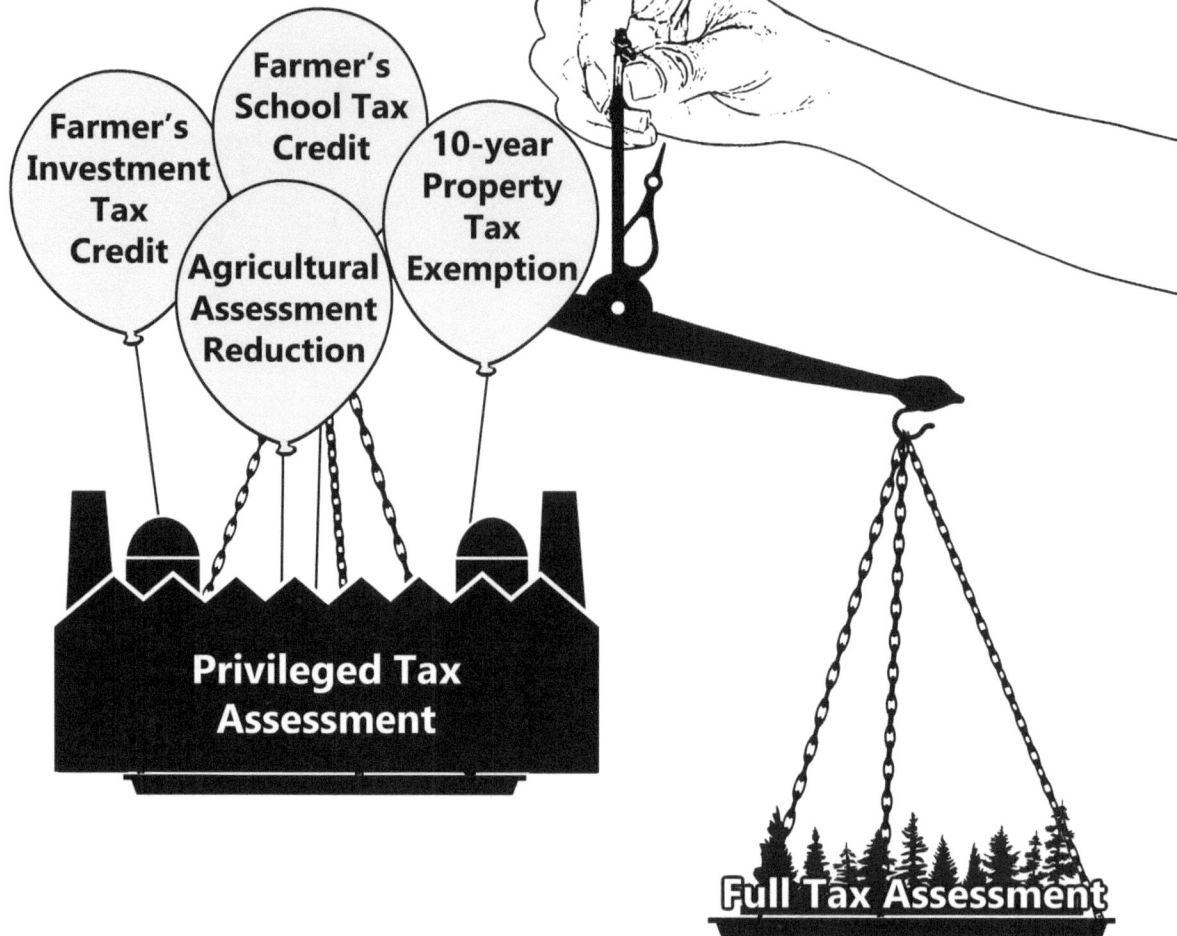

You know you live near a factory farm,
when you pay more property taxes
for your woodlot than they do for their farm.

You know you live near a factory farm,
when you're always being told
how important they are.

You know you live near a factory farm,
when your kids go fishing
with a pool skimmer.

You know you live near a factory farm,
when "manure spreading" means
it's spilled into the river again.

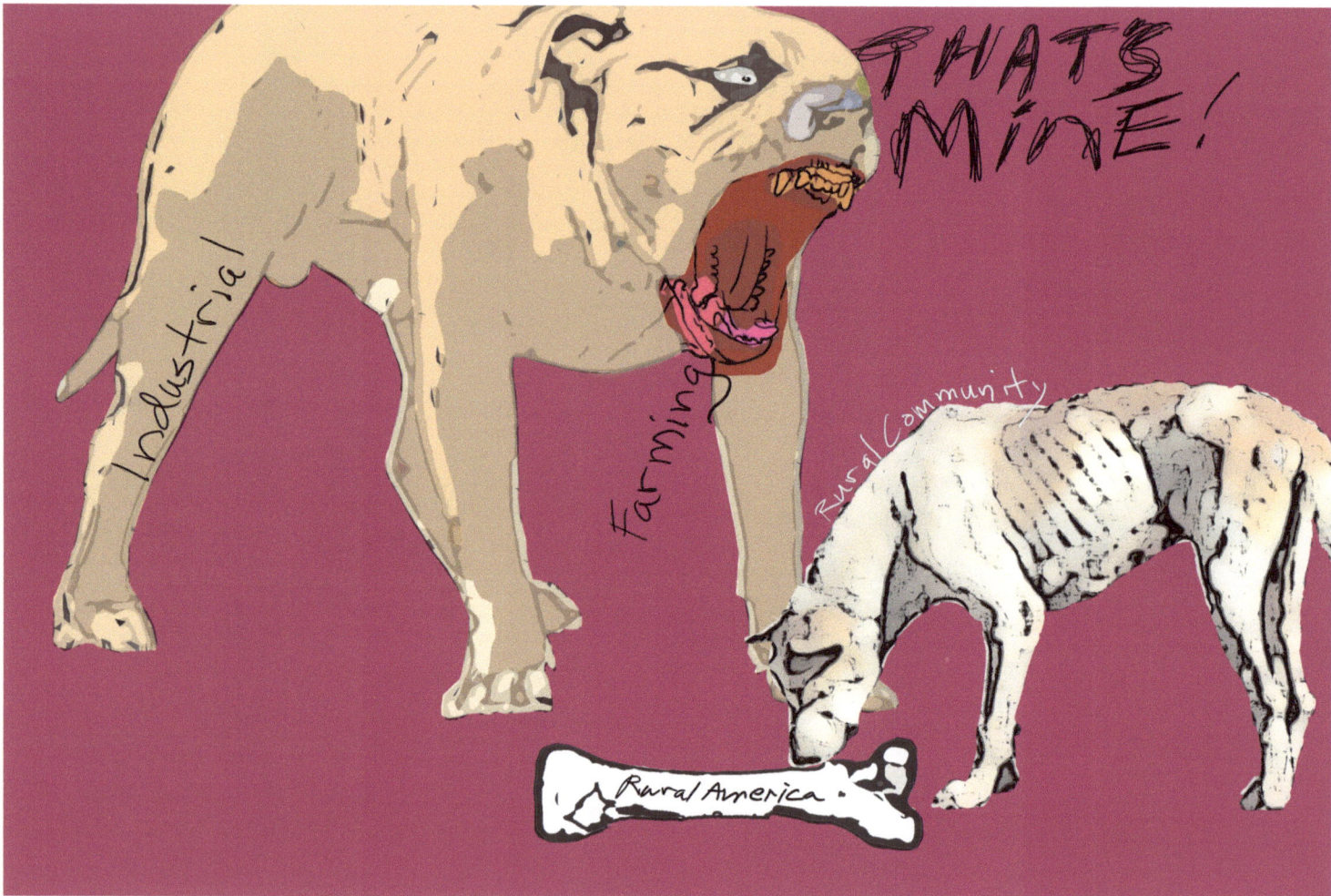

You know you live near a factory farm,
when their community vision is
"the Big Dog rules."

You know you live near a factory farm,
when your taxes subsidize their workers
and their workers send it out of the country.

You know you live near a factory farm,
when blowflies are the top
of the food chain.

You know you live near a factory farm, when the pitter-patter on your roof isn't rain.

That's a picture of your father fishing in Creasy Hollow when he was a boy — yes, there were fish back then . . . and people went swimming.

You know you live near a factory farm, when the grandparents have to tell your kids what nature used to be like.

You know you live near a factory farm,
when you're a long way
from a country club.

You know you live near a factory farm,
when they buy more antibiotics
than a hospital.

You know you live near a factory farm, when your brother's sick from bacteria no one's ever seen before.

You know you live near a factory farm,
when "agritourism" means
they've come to study a new pathogen.

"Please select from the following menu: 1. We're away from our desk right now. 2. We're in a meeting. 3. All our personnel are busy. 4. Your call is important to us — Please hang up and try again. To repeat this menu . . ."

You know you live near a factory farm,
when county employees
won't pick up the phone.

Your siding may be melting, but these are "fugitive gasses." And it's legal.

Case Closed!

DEC

You know you live near a factory farm, when the air you breathe would get a toxic waste dump shut down.

Dairy Cows

• **The life expectancy of the average cow in natural conditions is 25-30 years**

• **On the typical factory farm they live only four to five years**

Factory Farm Neighbors

Neighbors experience long-term exposure to the following environmental hazards:

• *Exposure to herbicide drifts and overspray [many herbicides are human carcinogens]*

• *Exposure to 168 gases from factory farm waste [including hydrogen sulfide that can cause irreversible brain damage and death]*

• *Exposure to nitrogen and phosphorus in well water from liquid manure spreading [nitrates in drinking water can cause "blue baby syndrome", a potentially fatal blood disorder]*

• *Exposure to high levels of E. coli and fecal pathogens in surface and drinking water*

• *Exposure to particulate matter and other pollutants linked to chronic respiratory ailments including bronchitis and obstructive pulmonary disease.*

While scientific papers cite the need to investigate the adverse health effects of factory farms on nearby residents, no comprehensive or long term studies have ever been undertaken.

Factory farm owners are pushing states to pass laws that would exempt them from any liability for the effect of their activities on the health of rural neighbors — states are complying.

You know you live near a factory farm, when their animals and their neighbors both have a reduced life expectancy.

I hear Jim got it too.

He called them and was angry about it being carcinogenic — all the spraying around his house.

But he couldn't do anything

The Bakers sold their land and moved away — now the factory farm owns it, right down to the creek.

Well, the _
  Mr. Richards?
  The test results are back.

You know you live near a factory farm,
when everyone you know
is getting cancer.

You know you live near a factory farm,
when your viewscape
is turkey vultures circling the offal bins.

Deep Plowing

• **Native grasses displaced**

Dust Bowl — South Dakota

1930s Dust Bowl

Today's Industrial Farming is responsible for:

• **70% of our water pollution**

• **80% of our antibiotics use**

• **Depletion of our aquifers**

• **Harmful Agal Blooms**

• **Methane driven Global Warming**

• **8,000 square mile Dead Zone in the Gulf of Mexico**

You know you live near a factory farm, when their farming practices are more disastrous than those of a century ago.

Big Agriculture: "Good Boy!"

Positive Findings

Career Longevity

Agricultural Research

You know you live near a factory farm, when agricultural studies always produce the desired result.

and we can't rule out Extraterrestrials as the cause.

14:17 / 2:08:53

You know you live near a factory farm, when politicians and bureaucrats make excuses for them.

We need to try to get to a point where we reduce the frequency and duration of blooms and restore some of the uses of these lakes.

Why not stop the farm runoff and restore all the uses?

While it's true that farming is responsible for more than 80% of the nutrient pollution, we've created new studies to show that we still can't be certain what is causing these agal blooms.

Regulating agriculture just doesn't make good political sense.

You know you live near a factory farm, when the public is educated to accept the pollution, not stop it.

You know you live near a factory farm,
when the snow-melt
is brown.

[To the tune of "Green Acres"]

Green lakes is just the place for me.
Rich farms and rural poverty.
Shit spreadin' out so far and wide.
Corporate farms destroying the countryside.
New York is where those farmers stay.
Big subsidies that sure ain't hay.
They just adore the green lake view.
All that rural land for the favored few.
[Rural] . . . Bad air.
[DEC] . . . Don't care.
[Rural] . . . My wife!
[Politician] . . . That's life.
Good bye to my wife.
Good bye, rural life.
Green lakes, now we are there.

You know you live near a factory farm,
when you dream of
a new sitcom.

You know you live near a factory farm,
when you have to
pressure-wash your cat.

You know you live near a factory farm,
when it's
no laughing matter.

You know you live near a factory farm,
when politicians won't kiss
your blue baby.

The End

About the Author

Doug Baird is an artist and writer living in Lansing, New York, who believes that both art and humor have transcendent properties.

Doug is project leader for the *Idea Enhancement Project*, a fiscally sponsored project of the New York Foundation for the Arts, exploring the use of art as a practical tool for increasing innovative and creative thinking.

His blog, *Rural Tompkins County — The Road to Hell is Paved with Good Credentials*, investigates elitist policy making in New York, and its effect on the rural community.

He is the author of two poetry collections: *As a Poet, I have a Confession* and *Please Take Care when You Utter a Curse*.

Representative artworks can be viewed at DougBairdArt.com

www.ingramcontent.com/pod-product-compliance
Lightning Source LLC
Chambersburg PA
CBHW060833270326
41933CB00002B/68